ICME-13 Topical Surveys

Series editor

Gabriele Kaiser, Faculty of Education, University of Hamburg, Hamburg, Germany

W0081775

David Bressoud · Imène Ghedamsi
Victor Martinez-Luaces · Günter Törner

Teaching and Learning
of Calculus

 Springer Open

David Bressoud
Department of Mathematics, Statistics, and
  Computer Science
Macalester College
Saint Paul, MN
USA

Imène Ghedamsi
IPEIT
University of Tunis
Tunis
Tunisia

Victor Martinez-Luaces
FJR/Faculty of Engineering
University of the Republic
Montevideo
Uruguay

Günter Törner
Faculty of Mathematics
University of Duisburg-Essen
Essen, Nordrhein-Westfalen
Germany

ISSN 2366-5947   ISSN 2366-5955   (electronic)
ICME-13 Topical Surveys
ISBN 978-3-319-32974-1   ISBN 978-3-319-32975-8   (eBook)
DOI 10.1007/978-3-319-32975-8

Library of Congress Control Number: 2016939361

Printed on acid-free paper

This Springer imprint is published by Springer Nature
The registered company is Springer International Publishing AG Switzerland

Main Topics

- Main epistemological aspects of Calculus concepts;
- Calculus thinking and learning difficulties;
- Analysis of Calculus in the institutional context including classroom practices;
- Brief analysis of Calculus design in the research;
- Main aspects of the transition between secondary and tertiary education in Calculus.

Contents

Teaching and Learning of Calculus

1 Introduction

This "ICME-13 Topical Survey" aims to give a view of some of the main evolutions of the research in the field of learning and teaching Calculus, with a particular focus on established research topics associated to limit, derivative and integral. These evolutions are approached with regard to the main trends in the field of mathematics education such as cognitive development or task design.

The research in the field of Calculus education covers almost all of the general issues investigated in the area of mathematics education. The most important trend is related to Calculus design, which puts forward several considerations to build and to implement alternatives by taking into account the results of existing research in the whole field.

Specifically, this overview of research includes a description of the main theoretical frameworks used in the field of Calculus education; descriptions of punctual evolutions approached through the main trends in the field, with a particular attention to the concepts of limits, derivatives, and integrals; a description of the state of Calculus instruction from both the European and American perspectives; a brief summary of the research progress and some new issues initiated by this progress.

As a complement to the main text, an extended bibliography with some of the most important references about this topic is included.

© The Author(s) 2016
D. Bressoud et al., *Teaching and Learning of Calculus*,
ICME-13 Topical Surveys, DOI 10.1007/978-3-319-32975-8_1

2 Survey on State-of-the-Art

2.1 Theoretical Frameworks

This section aims to give a global vision of research on learning and teaching Calculus, we use the main issues investigated in recent research as a filter in order to structure it. Much of this research deploys constructs from well-defined theoretical frameworks in the field of mathematics education; others act from a more empirical point of view.

Research dealing with cognitive development is based essentially on the frames of Concept Image and Concept Definition (CID) of Vinner and Hershkowitz (1980) (see also Tall and Vinner 1981; EMS 2014), on the theory of Register Semiotic Representation (RSR) of Duval (1995), on the Action-Process-Object-Schema (APOS) theory of Dubinsky (1991) and on the Three Worlds of Mathematics (TWM) of Tall (2004). Specifically, the CID framework highlights the distinction between the mathematical concept as formally defined and the individual's total cognitive representation for that concept. In this spirit, the concept image is the total cognitive structure associated to the concept containing all the mental pictures and related properties and processes, including conceptions and structural elements. A part of a concept image, the concept definition, is associated to the individual's definition(s) of the concept, learned by rote or self constructed (Vinner 1991). The RSR theory of Duval (1995) focuses on the mental relationships that exist for the individual between signifiers structured into semiotic registers and that which is signified (mathematical concept). In this theory, the constructs of *treatment* (leading to different representations of the concept in the same semiotic register), and *conversion* (different representations in different registers) are used to analyze the efficiency of tasks regarding mathematical concepts. The APOS theory is a model of learning mathematics with four stages based on Piaget's theory of constructivism with a main focus on assimilation and accommodation. Using Piaget's reflective abstraction (Dubinsky and McDonald 2001), this model describes the transitions between four stages from perceiving a mathematical concept through actions until regarding it as an object formed by encapsulation of the process. TWM is a theory involving three stages in the model of mathematical concept formation relating to three mathematical worlds: the conceptual-embodied, the proceptual-symbolic, and the formal-axiomatic. In the first world, which is applied to mathematical concepts perceived by the senses, we construct mental conceptions of the concept by using physical perceptions. In the proceptual world, the actions on the mental conceptions become encapsulated through the use of symbols. In this sense the term procept stresses the existence of a dialectic between process and concept as the same symbol can both evoke a process and the concept produced by this process. The transition to the formal world requires going ahead of the procepts to the formal definitions, through formal thinking or natural thinking (Tall 2008). Contrary to APOS theory, the TWM is more flexible since it does not emphasize a particular significance to the order of learning levels.

This framework is used to formulate the growth of ideas in calculus including two significant discontinuities: the shift from finite processes in arithmetic and algebra to the potentially infinite limit concept and the shift from embodied thought experiments and symbolic calculations to quantified definitions and proofs.

Research using theories with socio-cultural, institutional and discursive perspectives are generally based on commonly used approaches; these are the cases of the Theory of Didactic Situation (TDS) of Brousseau (1997), the Anthropological Theory of Didactics (ATD) of Chevallard (1985) and the Commognitive Framework (COF) of Sfard (2008). Some of the research related to the designing of tasks on learning and teaching Calculus use main tenets of TDS. The central object of TDS is the notion of Situation, which is defined as the ideal model of the system of relationships between students, a teacher, and a mathematical milieu. The learning process is highlighted through the interactions taking place within such a system. In the Situation, the students' work is modelled at several levels with a main focus on the action, on the formulation via the building of an appropriate language and on the validation using a coherent body of knowledge (González-Martín et al. 2014). The students' work grows up within and against a mathematical milieu during the phases of action, formulation and validation by optimising the interactions with peers and the teacher. The TDS constructs are also used to analyse regular Calculus courses, in ways that the interactions taking place within the system formed by the teacher, the students and the milieu are governed by the actual didactic contract and evolve according to its nature. In the field of learning and teaching Calculus, the ATD constructs are generally used to compare several institutional contexts, as well as to describe the organisation of mathematical activities related to a concept in one institution. To model mathematical activity, ATD uses the notion of praxeology, which is a complex, formed of two blocks: practical and theoretical. The practical block contains types of tasks and techniques for solving tasks. The technologies (namely the discourses which justify used techniques) and the theory that structure technologies are the components of the theoretical block. By modelling mathematical activities related to Calculus concepts in mathematical praxeologies, researchers describe what could be taught and what was taught (Winslow et al. 2014). Other research uses the COF to analyse the discourse of students and the discourse of teachers, as well as the mathematical communication between students and teacher. The basic assumption of COF is that "*learning mathematics is initiation into the discourses of mathematics involving substantial discursive shifts for learners—and the teaching of mathematics involves the facilitating of these shifts*" (Nardi et al. 2014, p. 184). Mathematical discourse can be specified according to: used words; visual mediators; endorsed narratives and routines such as defining, conjecturing, proving and so on. Processes, such as the production of proof, through which we become sure that a narrative can be endorsed, is called substantiation of a narrative. In the COF, mathematical communication involves continual transitions from signifiers (words or symbols that function as a noun) to realisation of the signifiers, which is a *perceptually accessible entity so that every endorsed narrative about the signifier can be translated according to well-defined rules into an endorsed narrative about the realization.* For instance, the signifier 'function' leads to several realisations such as an algebraic formula or a graph or a table of values. A realisation can be further realised, in which case the former realisation becomes a signifier for the later. In these senses, mathematical communication depends highly on the interlocutors' understanding of signifiers and can lead to a commognitive conflict that is not always detected by interlocutors. According to COF, this conflict is implicitly resolved by a mutual adjusting of interlocutors' discursive ways.

2.2 Potential and/or Actual Infinity: Beyond the Status Quo

Much of the research focusing on epistemological aspects of Calculus concepts underline the complexity of the switch from infinitesimal Calculus to formal Calculus. Based on both differential and integral Calculus, infinitesimal Calculus has been growing up over more than two thousand years. For these studies, the main issue concerns the cornerstone concept of limit involving infinitesimals and infinity. This notion shaped the contemporary approach of core Calculus concepts in standard analysis for real numbers, sequence convergence, series convergence, derivative, integral, differential and so on.

Historians commonly attribute the origin of rigorous Calculus to Cauchy, initiated in his *Cours D'Analyse* of 1821. In recent years, there has been research claiming to qualify Cauchy's contribution to the development of the modern view of Calculus. These studies led to more specific questions about the learning process of Calculus and its link to the approach used to introduce Calculus concepts. Before addressing these studies in more detail, it becomes important to examine research dealing with the potential schism between infinitesimal process and the modern standard limit notion, known to have started with the work of both Cantor and Weierstrass.

In her epistemological study concerning the limit concept, Sierpinska (1985, 1990) identified epistemological obstacles (Brousseau 1983) that are highly connected to infinitesimals and infinity. She claimed that overcoming these obstacles is synonymous with understanding. For example, perceiving a sequence as a long list of numbers that never ends, as well as believing that the meaning of terms like "approach" or "tend to" depends on the context are conceptions that may function as obstacles to thinking faithfully about limit. In the same vein, Kidron and Tall (2015) argued for more consideration, in teaching Calculus, of the distinction between potential and actual infinity of the limit process. Particularly, they showed a subtle parallel between mathematicians' work with the limit process before Cauchy and students' conceptions of limit. For students, limit is often seen in terms of the potential infinity of the on-going process rather than the fixed limit that can be calculated to any desired accuracy. Bagni (2005a, 2007) corroborated these results. In his research concerning Gandi's infinite series, he pointed out analogies between students' justifications and proofs of eminent mathematicians from history such as Leibniz and Euler. Similarly, Tall and Katz (2014) argued that some students' conceptions of sequence convergence can be found in historical understandings. According to them, conceptions such as "a sequence does not reach limit" and "a sequence can cluster around one or more points" are respectively consonant with the understandings of Newton and Cauchy. Assuming that these parallels should not be stated uncritically, Bagni (2005b) went further, asking for the investigation of the historical development of representation registers. According to him, this type of study permits anticipation of student difficulties and the creation of alternative historical situations based on kinetic and epsilontic versions of limit with an appropriate focus on employing semiotic registers. But, in what ways did mathematicians work historically with limit process and/or did they?

Błaszczyk et al. (2013) made clear how difficult it is to decide on this issue, knowing that several interpretations are usually encompassed by researchers' cultural backgrounds. Using several lenses, they showed that in the field of Calculus, claims referring to *the anachronistic idea of the history of analysis as a relentless march toward the yawning heights of epsilontics* (p. 46) needed debunking. It is widely acknowledged that the development of the history of analysis led at least to two modern versions strongly linked to the structure of the mathematics continuum: standard analysis anticipated by the epsilontic limit definition of Weierstrass (in the context of an Archimedean continuum) and non-standard analysis based on infinitesimal-enriched continuum which includes not only infinitesimals but also their inverses. As Błaszczyk et al. (2013) did, Borovik and Katz (2012) and Tall and Katz (2014) argued that it is an oversimplification to interpret Cauchy's work according to an Archimedean context. They exposed definitions stated by Cauchy, including that of continuity, intermediate value theorem and the summation of series, to show that infinitesimals actually have not been eliminated in his contributions. The issues concerning the implications of such studies in the field of mathematics education are still topics of interest. The investigation of such issues should be done by taking into account multiple relationships between the principal actors of the mathematics education system including, the mathematics itself, students and teachers.

2.3 Students' Difficulties: Dealing with Informal Calculus Thinking

Different areas of research dealing with cognitive development have at least one feature in common; they were planned to investigate students' thinking about Calculus concepts. In all cases, the central project of these studies is closely associated to the issue of students' conceptions (concept image in the terms of COD framework) and their evolution. Many of these conceptions are rooted in the student's previous experience. The inefficiency of these conceptions is described by Cornu (1991) as *cognitive obstacles*, a term introduced by Bachelard (1938). Using COD and TWM frameworks lenses, many of these studies built their analysis on individual student work. More recently, some studies have drawn on the COF framework to investigate communication in the context of small groups of students discussing Calculus concepts.

2.3.1 Limits

One piece of pioneering research concerning limit has emphasized the difficulty for students to conceive the limit process as a number (Tall and Vinner 1981). Cottrill et al. (1996) observed that the concept of limit of a function actually encapsulates two processes: one in which the independent variable approaches a value and one in which the dependent variable approaches a value. According to these researchers, the limit concept is usually stated by students as a dynamic process; almost all students' images are organised according to this idea. Tall and Vinner (1981) have demonstrated the distinction in observing students who will accept that 1 is the limit of the sequence 0.9, 0.99, 0.999, ... while still asserting that 0.999..., which they understand to be an encoding of this process, does not *equal* 1 (Tall and Schwarzenberger 1978). One of the problems with student understanding of limit, as pointed out by Cornu (1991) and others (Cottrill et al. 1996) is that the limit as process is *unencapsulated*. Orton (1980), Davis and Vinner (1986), Cornu (1991), Tall (1992), and Oehrtman (2009) have explored the ways in which students deal with nonterminating processes and have recognized that they often draw on their experiences of finite but long-running processes. This can lead to what Tall (1992) has referred to as the *generic limit property*, the assumption that, as with the last of a finite sequence of steps, any properties possessed by all of the intermediate steps must also hold for the limit. Thus the limit of 0.9, 0.99, 0.999, ... must be strictly less than 1 because all of the intermediate values are less than 1. Similarly, it leads to the student expectation that an infinite sum of continuous functions must be continuous.

All the phenomena described above have the same main source: conceptions that students draw from their ideas, intuitions, and knowledge in trying to make sense, but that can block understanding. Among those related to the dynamic interpretation of the limit (Tall and Vinner 1981; Robert 1982; Williams 1991; Cottrill et al. 1996), one comes from the commonly used language of "approaching". As Tall and Vinner (1981) and Robert (1982) have documented, the word "approach" can create a counterproductive concept image, implying monotonicity as well as creating or reinforcing the belief that the function will never equal the limit. Even more problematic is that "approach" describes an action or process, placing the concept of limit on the bottommost rungs of Dubinsky's APOS theory (Dubinsky and McDonald 2001). However, Williams (1991) and Oehrtman (2009) argued that this dynamic interpretation is usually conceived not as continuous but as sequential, *"an idealized form of evaluating the function at a series of points successively closer to a given value"* (Williams 1991, p. 230). This creates cognitive dissonance when students try to combine their understanding of limit as a process with the assertion that the limit is a value.

Oehrtman (2009) built on Black's (1962, 1977) theory of metaphorical attribution to describe in considerable detail one of the mechanisms employed by students who embrace a distinction between the limit as process and the limit as value. He calls this the *collapse metaphor*. He repeatedly observed students who interpreted the limit process as getting closer and closer until at some point this process collapses onto the limit value. This collapse metaphor is also implicit in earlier work of Thompson (1994) who observed students who would explain the second fundamental theorem of Calculus (the derivative of a function that is defined by a definitive integral) by setting up the limit definition of the derivative, then ignoring the denominator and explaining that as the change in x approaches 0, the integral, taken over shorter and shorter intervals, collapses down to just the value of the function. In his study, Oehrtman (2009) catalogued the common metaphors employed by students as they attempt to explain the meaning of statements involving limits. As with concept image, most students will employ a variety of metaphors that are often applied in a manner that is context specific. Among the most common metaphors are *proximity* (when x and y are close, $f(x)$ and $f(y)$ will be close), *infinity as number* (using infinity as if it is a very large number), *physical limitation* (assumption that there is a smallest positive number), *collapse*, and *approximation* (the metaphor that comes very close to and may actually embrace the formal definition of limit).

Concept images of limit may be fluid, but, as many researchers have documented (Sierpinska 1987; Williams 1991; Szydlik 2000; Oehrtman 2003; Przenioslo 2004; Roh 2008), they are also difficult to overcome. Specifically, dynamic images based on operational conceptions (Sfard 1991) are resistant and could seriously prevent students from achieving an appropriate understanding of the formal definition of limit. Building on this, Przenioslo (2004) studied the development of students' images of limit concept after being exposed to formal definitions. She pointed out the emergence of an image of neighbourhoods and the obstinacy of dynamic images among students who continue to think about limit by referring to *graph approaching* or *values approaching*. This result is sustained by the study of Roh (2008), which showed that students' understanding of the limit concept, as they expressed it through their individual definitions, is conditioned by their dynamic images of limits as asymptotes (without ever reaching it) or cluster points. These images could be built on conceptions that arise in every day usage of the word "limit" as a boundary (Davis and Vinner 1986; Williams 1991; Petterson and Scheja 2008; Oehrtman 2009). This can be manifested in confusion between accumulation points and limits, especially if the accumulation points also serve as upper and lower bounds on the sequence, thus allowing a sequence to have more than one limit (Roh 2008; Mamona-Downs 2010). It can also appear in the belief that the function cannot assume the limiting value in any punctured neighborhood (Davis and Vinner 1986; Przenioslo 2004; Nair 2010).

As highlighted by Martin (2013), students' concept images could also be founded on structural conceptions as well as pseudo structural ones (namely conceptions which are *both partial in operation and incomplete in structure*). In his study of Taylor series convergence, these images are directly related to elements of the mathematical structure and the potential operations on those elements. These elements include the use of particular values for the independent variable, the work with terms, the focus on partial sums, and the work with remainders. According to this study, students' images are constructed on pseudo structural conceptions that generate difficulties in moving between these images for effective engagement with the given tasks. It appears that the mathematical structure of the limit concept is another source of influence on the understanding of definitions of several Calculus concepts related to limit.

Several studies have explored student difficulties with the formal epsilon-delta definition of the limit of a function (Tall and Vinner 1981; Swinyard 2011; Swinyard and Larsen 2012; Bezuidenhout 2010; Oehrtman et al. 2014), demonstrating that students often focus on key phrases rather than their logical connection, thus accepting nonsensical combinations of these phrases as legitimate while rejecting correct definitions cast in different phrases. Hardy (2009) has documented that many of the students whose work with limits has been restricted to finding limits of rational functions emerge with no understanding of what they are doing. The acquisition of quantified statements is a complex process that can not be produced simply from concept definitions (Cornu 1991). In the case of the limit concept, students may construct their own meaning of the logical definition that could be by a long way far from its conventional sense (Davis and Vinner 1986). In her analysis of the role of symbols in the formal definition of sequence convergence, Mamona-Downs (2001) argued that the focal compound of this quantified statement is the one related to the inequality measuring the closeness of sequence terms to the limit with ε error; the focus will then be in the variability that each symbol of this inequality has. Roh (2010a) synthesized the complexity of the work with such variability through the relationship between ε and N. This relation is shaped by the arbitrariness of the error bounds decreasing towards zero. In the case of the least upper bound concept (supremum), Chellougui (2009) argued that students are unable to perceive easily why quantification should be in one order more than in another. In the same spirit, Swinyard (2011) discovered that one of the greatest obstacles to the correct mathematical understanding of limit of function was student preference to focus first on the change in the independent variable, understanding the limit as x approaches c of $f(x)$ to mean that as x gets closer to c, $f(x)$ gets closer to the limiting value, a formulation that can be useful for finding limits but that is rife with opportunities for misleading concept images. Furthermore, Swinyard is representative of a growing body of research into how students use their understanding of limits to reason about limits (Swinyard and Larsen 2012). According to Swinyard (2011), students are able to reinvent a coherent definition of limit of function with a high level of significance. This study provides a detail account of how students might think about limit formally. These details are encapsulated in an exploratory model of several levels of students understanding (Swinyard and Larsen 2012).

The aforementioned studies carefully explain the difficulties of students in achieving a higher understanding of the formal definition of limit. The degree of legitimacy of this ultimate accomplishment is largely discussed in Vinner (1991). For Alcock and Simpson (2011), students do not necessarily feel the need for classifying a mathematical object as a member of a coherent set in which elements obey the same formal definition. They argue that before deciding on the use of a definition by students, it is important to investigate what they call *concept consistency* in order to consider student performance in the judgement of classification.

In all cases, in the terms of TWM framework, the difficulties mentioned above are embodied and symbolized in previous experiments. As acknowledged, the necessity to create a rigorous understanding of infinite processes and more general Calculus concepts initiated the shift in formal mathematics to the use of quantified statements without a natural base.

The process-object duality of Sfard (1991) has been the subject of experiments by Bergé (2010) relating to completeness. This study shows that students have difficulties in linking the operational perspective of completeness and the supremum of subsets to the structural concept of real numbers. Based on COF constructs, Kim et al. (2012) point out that in the case of the notion of infinity, student understandings are highly correlated to their use of words and visual mediators. According to these researchers, the analysis of the discourses on infinity of Korean and English students can lead to their categorisation in terms of process and object point of views. This study shows that Korean students' approach to infinity is more structural and formal as opposed to the English one, which is apparently more procedural and informal.

2.3.2 Functions

Vandebrouck (2011) addressed the evolution of function thinking from secondary school to university; he strengthened the difficulties for students to transit from a pointwise and global perspective on functions to a more local perspective as required by the formal Calculus world of the university. According to this study, this transition requires a complex conceptualisation of the function concept in terms of process and object duality. This conceptualization required an early start for the development of the variational thinking in students (Warren 2005; Dooley 2009; Warren et al. 2013).

Using the dialogue between APOS and RSR frameworks, Trigueros and Martínez-Planell (2015) focused on student graphical understanding of two variable functions. The results stressed the instructional consistency of the activities in the classroom.

2.3.3 Derivatives

Orton (1983b) provides one of the earliest descriptions of student difficulties with derivatives. While the students he studied were generally proficient at computing derivatives, he found significant misunderstandings of the derivative as a rate of change and of the graphical representation of the derivative. These were frequently tied to a poor or inadequate grasp of limits as well as ratio and proportionality. Byerley et al. (2012) have studied student difficulty in recognizing the quotient as a measure of relative size and shown how this aspect of ratio and proportion impedes student understanding of the derivative.

Ferrini-Mundy and Graham (1994) used interview methods to document student difficulty in connecting the symbolic representation of a derivative with any kind of geometric understanding. Confirming what Orton had found, they discovered that students could compute derivatives while being unable to connect those results to such tasks as the production of the equation of a tangent line. This work was expanded by Nemirovsky and Rubin (1992) who both described and explained student difficulties in connecting features of the graph of a function with its derivative. A student's natural tendency is to associate features of the graph of a function with features of its derivative. Thus, students will often identify the value at which a function is maximized with the value at which the derivative is maximized. Additional work on student difficulties linking the graph of a function with the graph of its derivative has been done by Borgen and Manu (2002).

Aspinwall et al. (1997) have described how, even for a student who fully understands the derivative as the slope of the tangent, sketching the graph of the derivative from the graph of the function can be impeded by previously encountered images. Bingolbali et al. (2007) showed the influence of the departmental affiliation of students on their developing conceptions of the derivative. The findings reveal that mechanical engineering students develop a tendency to focus on rate of change while mathematics students develop an inclination towards tangent-oriented aspects. He suggested that departmental affiliation appears to have an influence on cognition and plays a crucial role in the emergence of different tendencies between the two groups.

Asiala et al. (2001) used the framework of APOS theory to analyze student attempts to create the graph of the derivative from either the graph of the function or from a list of inequalities satisfied by the function, the derivative, and the second derivative. They were able to identify student difficulties as an instance of an inadequate understanding of function as a process. Along similar lines, Confrey and Smith (1994) and Thompson (1995) have located many of the problems with interpretation of derivatives within the common understanding of a function as a static object rather than as a description connecting two covarying quantities. White and Mitchelmore (1996) documented the difficulties encountered by students who approach derivative problems by focusing on the variable as a symbol to be manipulated rather than understanding it as representing a varying quantity. Recently, Thompson and Carlson (in press) have elaborated on the importance of covariational reasoning as critical to developing a conceptual understanding of the derivative.

Baker et al. (2000) extended the work of Asiala et al. emphasizing student difficulty in incorporating knowledge about the second derivative into producing the graph of the function, thus highlighting how hard it is for many students to coordinate all of the information that may be available. The authors conjecture that this may arise from a lack of recognition of the derivative as a function.

In 2000, Zandieh offered a framework for understanding student difficulties with the concept of derivative, emphasizing the need for multiple representations or contexts as well as layers of process-object pairs. An important insight in this paper is the recognition that students often employ pseudo-objects, an object that is not accompanied by a structural understanding of an underlying process. Sometimes, this is not a hindrance. Thus, a student can know that the derivative can be used to find speed without understanding the limit process that explains why this is the case. However, this lack of understanding can lead to misapplication of the derivative or inappropriate interpretation of its result. Zandieh explained the difficulty of moving from the notion of derivative at a point to derivative as a function as an example of understanding the relationship between a higher process-object layer (derivative as a function) to the one on which it rests (derivative at a point). Others, including Habre and Abboud (2006), have built upon this framework. Using COF constructs of *use of words* and *visual mediators,* Park (2013) explored students' discourse about the derivative as a function based on the concepts of function at a point and function on an interval. This researcher argued that the common description by the students of the derivative as a tangent line is linked to their use of the word "derivative" for both the derivative of the function and the derivative at a point.

2.3.4 Integrals

Kirsch (1976) made one of the earliest contributions to the understanding of how to approach integration. He particularly highlighted the role of the visualization of the fundamental theorem of calculus arguing that a conceptualization of the relationships between the integral and the derivative could be achieved through an appropriate perception of the derivative as both a rate of change and a slope of a tangent line.

Rasslan and Tall (2002) describe an experiment in which 41 secondary school Calculus students were asked to define the definite integral. Their responses fell into five categories: definite integral as area ($n = 4$), definite integral as the difference of the antiderivative evaluated at the endpoints ($n = 3$), via an example of an actual definite integral calculation ($n = 3$), an erroneous statement such as the change in the value of the function to be integrated ($n = 5$), or no answer ($n = 26$). No one defined it as a limit or made reference to a Riemann sum. The authors question the traditional approach to integration, which begins by approximating areas, then derives rules for computing areas under polynomials, and then introduces the Fundamental Theorem of Calculus as the tool that makes it easy to find areas. Grundmeier et al. (2006) have demonstrated that an understanding of integral as area does very little to help with student ability to use definite integrals. Sealey (2006) has also documented student difficulty connecting the concept of definite integral as area with definite integral as accumulation. In his study related to integral concept in the first year university, Haddad (2013) categorised students' difficulties regarding the distinction between area, antiderivative and integral. He argued that the inefficiency of students' knowledge is linked to an established confusion between these notions.

Perhaps because of this, students have a great deal of difficulty in understanding the definite integral as a limit of a sum. Orton (1983a) observed this and suggested that the root of the problem lies with student difficulties with limits in general. This point has been elaborated by Sealey and Oehrtman (2007) who investigated the role of approximation as a vehicle for improving student understanding the definite integral as a limit.

Thompson and Silverman (2008) have investigated student difficulties with the concept of integration as accumulation. They point out that most students fail to recognize the Riemann sum as an encapsulation of an accumulation process. As Thompson (1994) has shown, this can be a serious impediment to understanding the Fundamental Theorem of Calculus. Carlson et al. (2003) have demonstrated that an emphasis on covariation can assist the development of student understanding of accumulation and its relationship to the Fundamental Theorem of Calculus.

All these researchers examine students' work with no more associations with the instructional context. Studies discussed in the following have that perspective.

2.4 Analysis of Calculus Curricula

2.4.1 Textbooks and Calculus Materials

Most of the research studying mathematical textbooks and materials deploy the constructs of the ATD framework to model the mathematical organization related to the underlying Calculus concept. Some of these studies combined several frameworks (ATD, TDS, TWM and RSR) for more insights.

Among these studies, those that investigate the transition between secondary and tertiary education in Calculus assume that the mathematical cultures of secondary and tertiary institutions are the main foci of students' difficulties. It is obvious that this basis axiom does not contradict the results of studies focusing on cognitive development. The study of Praslon (2000) concerning derivative in France is a pioneer work in that field. Based on the categorization of mathematical praxeologies in the environment of derivative in both secondary school and first-year university, this study showed that the secondary-tertiary transition in the derivative environment is not a simple transition between embodied/proceptual worlds and the formal world. This transition is a blend of changes in several dimensions related to instructional expectations in the two institutions: process-object duality, particular-general objects, algorithmic-conceptual techniques, the choice of semiotic registers and the conversions between them, autonomy given to solve problems and routinization of practices, diversification of problems, etc. Building on this, Bloch and Ghedamsi (2005) focus on the crucial differences in the environment of the limit between the end of secondary school and the first-year university in Tunisia. The critical analysis of textbooks and Calculus materials led to the categorization and the formalization of many important changes that should occur in the way students are required to work at the first-year university. This study deploys TDS constructs of didactical variables (namely the parameters that influence the mathematics students' work) to characterize these changes. Some of these didactical variables highlighted the role of the formalism, the use of proof settings, the use of technical methods, the level of operation of the notion (process-object duality, ability to be mobilized or not...), and the use of conversions between semiotic settings. In the case of the limit, the values given to the didactical variables from one institution to another are mutually exclusive. The use of this model addresses the gap between second and third level mathematics and does suggest some hints for how to design tasks in the transition. The Spanish study of Bosch et al. (2004) related to the concept of limit corroborated these results. This study shows the existence of strong discontinuities in the organization of mathematical praxeologies between secondary school and university, and builds specific tools for qualifying and quantifying these. In the same spirit, by means of the ATD framework Winsløw (2008) studies the transition from concrete to abstract perspectives in Denmark regarding the concept of function and the operations on these functions associated with the limit process. He argued that in secondary schools the focus is on practical-theoretical blocks of concrete analysis, while at university level the focus is on more complex praxeologies of concrete analysis and on abstract analysis. Using the same lenses, Bergé (2008) explored the organisation of mathematical praxeologies related to the set of real numbers and its completeness at the university level in Argentina. This study demonstrated that the completeness rests on both Calculus and analysis courses at the university; the difference can be shaped by what is expected by way of proof (the Theoretical block in ATD terms).

Another kind of ATD comparative study of the secondary-tertiary transition was undertaken by Dias et al. (2008) between Brazil and France on the topic of function concept. This study focused on the analysis of institutional relationships through the analysis of summative evaluations, used for the admission of secondary students at the university. By analyzing typical tasks in the two countries, they conclude that associated praxeologies were based on algebraic techniques and technology in Brazil while in France the associated praxeologies emphasized the use of analytic techniques and technology. Furthermore, the study also showed a higher level of students' guidance through hints and intermediate questions in France than in Brazil. The concept of function has been studied by Martinez-Sierra (2008). He focused on the study of the articulation of the algebra and trigonometry concepts with the elemental Calculus concepts, as the function concept.

One of his main results shows the conceptual breaks present in articulating concepts like the use of radian like angular measurement and the negative angles and angles larger than 360°.

González-Martín (2009) and González-Martín et al. (2011) combined principally RSR and ATD frameworks to analyze textbooks regarding infinite summations at the university level in Québec and UK. By focusing on the role of visualization in mathematical materials, this study showed few graphical representations and few opportunities to work across different registers (algebraic, graphical, verbal), few applications or intra-mathematical references to the concept's significance and few conceptually driven tasks that go beyond practicing with the application of convergence tests and prepare students for the complex topics that employ the concept of series is implicated.

The topic of differential equation has been investigated by Arslan (2005). He showed the dominance of an algebraic approach to teaching differential equations in the upper secondary school in France and the lack of numeric and qualitative study of ODEs. He formulated the hypothesis that the limitation to the algebraic frame only for the treatment of differential equations can be the origin of difficulties and habits which students face with qualitative interpretation tasks. Czocher et al. (2013) investigated topics in introductory differential equations in the US, and their relation with the knowledge that students are expected to retain from their Calculus courses.

2.4.2 Calculus Instruction in France[1]

Törner et al. (2014) provides a first overview of the landscape with respect to Calculus teaching in European classrooms, an area where research is very limited. In particular, they use a small expert-based survey and a literature review to trace the development of Calculus teaching at schools in a number of European countries and identify commonalities and differences. At the moment, Calculus instruction is not really under discussion (except for the continuing debate over the use of CAS), at least in Germany.

In France, the teaching of analysis officially begins in Première (Grade 11) and is closely associated with the teaching of functions and sequences. Functions are part of the program entitled "Organization and data management, functions" in Troisième (Grade 9), "Functions" in Seconde and finally "Analysis" in Première and Terminale (Grade 11 and 11). This instruction represents 40 % of the total curriculum (Grade 10 and 11) and 50 % in scientific and economic Terminale (Grade 12). Vandebrouck's studies (2011) have helped to identify three stages of development in the teaching of analysis (See also Kuzniak et al. 2015).

---

[1]Contributed by Alain Kuzniak, Université Paris Diderot-Paris 7.

First, from the end of Troisième (Grade 9) until the beginning of Première (Grade 11), students are expected to develop a form, F1, of work in analysis in which various and numerous semiotic representations of functions (including tables of variation, graphs and algebraic formulas) are used without any emphasis on algebraic formulas. This work is intended to enable students to conceptualize functions as a global object by coordinating the various registers of representations and connecting functions to other areas of mathematics, such as geometry, or other fields, such as physics or economics. There is thus a development of modeling activities supported by technological artifacts such as dynamic geometry software or spreadsheets.

From Première (Grade 11) to university, where the demands are more complex, a second form of work in analysis, F2, is developed and associated with algebraic operations and expressions. Local concepts are introduced gradually: limits, continuity, and differentiability. This last is introduced before the notion of limit. These local concepts are actualized primarily in problems in which functions are represented by algebraic formulas, usually polynomial, exponential, or logarithmic. An intuitive approach to the concept of limit appears with no formal definition. Thus, the "nombre dérivé" (derivative number) and the derivative function are introduced in Première (Grade 11). The "derivative number" is introduced as the limit of the rate of change $(f(a+h) - f(a))/h$ when h tends to 0, but no formal definition of the limit is given. Continuity is discussed in the same intuitive way in Grade 12. For example, it is said that "To indicate that u_n tends to l as n tends to infinity, we will say that any open interval containing l also contains all of the values of u_n from some subscript on". In fact, once these concepts have been introduced, no formal work is done on the objects, and most of the activities are devoted to calculation based on algebraic expressions. As noted by Coppé et al. (2007), the already large use of algebraic representations for the study of functions in Grade 10 textbooks (from 30 to 58 % depending on the book) becomes predominant in the Grade 11 and 12 textbooks.

The third form of work, F3, starts in the first year of university, and is particularly evident in lectures where formal proofs are given by university professors. This is the first discovery of the paradigm of infinitesimal analysis with no reference to algebra and with new rules such as those of quantification, which are now required. This paradigm shift requires new techniques and approaches: lower and upper bound techniques, the interplay between sufficient and necessary conditions with a local perspective on functions. This new form of work draws on traditional algebraic techniques as in simplifying algebraic expressions to find limits. New tasks using expressions like "close to" or "ever closer to" cannot be solved without the use of quantifiers to work on algebraic expressions. The foundations of this form F3 relate to the completeness of R under one of the three following forms: the convergence of increasing sequences with upper bound, the nested interval principle, or the convergence of Cauchy sequences.

All of the research in this domain shows that the transition between high school and university is very difficult in analysis. In high school, the work under the form F2 is relatively homogeneous and marked by the importance of algebraic expressions. Focusing on this form of work leads to an impoverishment of the possible interplay between different semiotic representations introduced in form F1. This impoverishment of the semiotic interplay is associated with a "routinized" work with algebraic techniques that limits students' initiative and gives them an erroneous conception of the very nature of the mathematical work in analysis, which seems to be reduced to some form of algebra.

At university, the focus is on a discursive validation supported by an important theoretical reference framework, for which students are not prepared in high school. Also, we note the disappearance of graphing calculators, which makes it difficult to use graphical representations that are not already familiar to the students. The desired mathematical development rests on the interplay between two different perspectives. It requires a complex treatment based on formal symbolic manipulations while assuming an understanding of how these relate to the former work on algebraic expressions. It is also worth noting that at university there is very little work on procedural fluency.

2.4.3 Calculus Instruction in Germany

Calculus in Germany is *the* traditional curriculum issue. At the beginning of the 20th century, it was Felix Klein who succeeded in making calculus compulsory in the curriculum for the upper secondary grades at Gymnasium at least from 1927. So, mathematics at the Senior High School was primarily calculus.

Secondarily, analytical geometry (including conics) was also taught, however calculus (differentiation and integration) remains the constant kernel with no change at a first glance with respect to the content for many decades. Of course, there was a rise in the discussion and the reflection in the '70s through the New Math movement and the influence of Bourbaki elements. There were some attempts to make calculus as rigorous as at the university level, (epsilon-delta-infinitesimal calculus), which were not successful. Calculus survived with a lessening of the content.

Today, the old (classical) analytical geometry has turned into a first introduction to linear algebra and stochastics has became compulsory at the Senior High School. However, calculus continues to dominate.

In the last twenty years software and new tools lead to a partial rethinking of the elements which should be taught, however, there is still a strong lobby within the teachers' communities and the editors of textbook not to change too much, thus discrete mathematics and stochastics are strongly opposed.

2.4.4 Calculus Instruction in the United States

Calculus in the United States is and always has been considered a university-level course. Curiously, it is now predominantly taught in high school. For most of those for whom Calculus is a prerequisite for an intended major, this same course is then retaken at university.

In the early 1950s, the College Board established the Advanced Placement (AP®) program as a mechanism for allowing high school students who are ready for university-level studies to do such work in their high schools and be certified, via a national examination, as having completed the full equivalent of a university-level course. Calculus was one of the first such courses. For the first two decades, only a small and elite group of students were able to take advantage of it.

Beginning in the late 1970s and accelerating through the 1980s, state education officers saw the introduction of Advanced Placement courses as a means of raising the quality of underperforming schools. They put in place generous financial incentives to include such courses in the school curriculum and channel students into them. Although there were serious attempts to prepare the high school faculty to teach these courses, the result was the proliferation of courses that were Calculus in name only. By 1986, the National Council of Teachers of Mathematics (NCTM) and the Mathematical Association of America (MAA) were sufficiently alarmed that they issued a joint statement warning of the dangers of accelerating students into a Calculus course for which they were not prepared or which bore little resemblance to the Calculus taught at university (Steen and Dossey 1986). A preliminary report of results from a large-scale survey of Calculus I students, highlighted students' mathematical background as well as aspects of instruction that contribute to successful programs.

But the floodgates had been opened. The presence of Calculus on a student's high school transcript is highly correlated with success at university. Admissions officers recognize this, and parents soon became aware that it had become a factor in admission decisions and the granting of financial aid. As more students enrolled in Calculus in high school, more parents pressured teachers and administrators to let their children in also. Growth rates for high school Calculus enrollment that consistently exceeded 13 % per year in the 1980s have slowly come down, but are still running close to 6 % per year (College Board 1997–2014). The result is that in 2014–2015, over 750,000 high school students were enrolled in Calculus, almost a quarter of all high school seniors in the United States. Over one-third of them then retake this course when they get to university (Bressoud et al. 2015).

In the U.S., virtually all university Calculus is taught within the mathematics department. At the major public universities, thousands of students are enrolled in Calculus each semester, providing much of the justification for a large mathematics faculty. Economies of scale are often achieved by using the same course for multiple constituencies. A typical first Calculus course will include students heading into the physical sciences and engineering as well as the life and social sciences, with the few mathematics majors added to the mix (Bressoud 2015). This is then taught in large lecture halls. The result is a course that is heavy on procedural fluency with little attention to conceptual understanding. In a national survey of Calculus I final examinations, Tallman et al. (2016) found that over 85 % of the test items involved a single step and could be solved by simple retrieval of rote knowledge.

One of the bright spots in U.S. Calculus instruction is a legacy of the Calculus Reform efforts of the early 1990s, the recognition in almost all textbooks and most universities of the importance of graphical and numerical in addition to algebraic representations of derivatives and integrals. Also, recent years have seen a great deal of experimentation: flipped classes, use of online resources, introduction of active learning approaches, and the development of courses that are targeted to specific disciplines such as biology and courses that are designed to meet the needs of students who have studied Calculus in high school but are not prepared to advance to the next course. It is now the rare mathematics department that is not trying or at least thinking of trying one of these ideas.

In their 2011 survey of 24 recognized American authorities on Calculus instruction, Sofronas et al. found that 17 identified *limit* as a central concept or skill. The AP Calculus Curriculum Framework (College Board 2015) identifies limit as one of the four big ideas of Calculus, the others being differentiation, integration and the fundamental theorem of Calculus, and series. This focus on limits reflects the perception that they are foundational to understanding Calculus (Ervynk 1981; Williams 1991). However, Sofronas et al. found that all 24 of their nationally recognized authorities on Calculus agreed that derivatives are a central concept. Two-thirds of them ($n = 16$) considered fluent ability to compute derivatives an essential skill. Half ($n = 12$) considered an understanding of the derivative as a rate of change to be central. Just under a third ($n = 7$) emphasized the importance of understanding the graphical representation of the derivative. Furthermore, almost all of the participants identified integration as a central concept of Calculus, and half listed the Fundamental Theorem of Calculus as a first-tier subgoal. The other first-tier subgoal was an understanding of integration, below which respondents listed "(a) the integral as net change or accumulated total change, (b) the integral as area, (c) techniques of integration".

2.4.5 Calculus Instruction in Uruguay

In Uruguay, Calculus instruction starts in the last year of Secondary School with the teaching of limits and derivatives. A few decades ago, it was usual to include other topics like integrals, sequences and series and Taylor polynomials, which nowadays are shifted to university first year courses. Something similar happened with the theoretical knowledge, which traditionally was evaluated in exams devoted to these topics. Nowadays, the exams are focused only on routine practical procedures such as calculating limits and derivatives, complemented with graphics of the involved functions.

Thus, those changes in the Secondary school content—and even more, the change in the mathematical level and maturity of the students who enter the university—had important consequences in the first year university Calculus courses. For instance, they start revisiting topics of Secondary school (like functions, limits and derivatives), which usually takes half a semester or even more. After that, a typical Calculus I course is completed with integrals and Taylor polynomials. Sometimes these topics are complemented with a few simple Differential

Equations—separation of variables and linear ODE—in most cases included in the syllabus to support Physics courses.

Taking all these facts into account, a typical Calculus I university course is expected to include:

- Derivative as the gradient of the tangent to the graph of a function at a given point
- Derivatives of polynomials, trigonometric, exponential and logarithmic functions
- Derivatives of sums, differences, products and quotients of functions
- Derivatives of composite functions
- Monotonic functions (increasing and decreasing functions)
- Stationary points (maximum, minimum and inflexion points)
- Use of second derivative test to discriminate between maxima and minima
- Finding equations of tangents to curves
- Integration as the reverse of differentiation
- Properties of integrals and main theorems
- Integration of polynomials, trigonometric, exponential and other simple functions
- Integration by parts
- Integration by substitution
- Integration of rational functions
- Improper integrals and convergence criteria
- Finding the area of a bounded region
- Finding the volume of revolution about the x- or y-axis
- Taylor and MacLaurin polynomials
- Sequences, Series and convergence
- Solving for the general solutions and particular solutions of differential equations

There are other issues that deserve to be considered. For instance, electronic calculators are not allowed in several faculties at UdelaR (the state public university) at least for first year mathematics courses. In most of the institutions students are allowed to consult their books, tables of derivatives and integrals, etc., during the exams, whereas in others they are expected to do their exams using only pencil and eraser.

Furthermore, in the last decades of the twentieth century, Calculus I exams used to be divided into two parts: one devoted to practical routine exercises and another one that was supposed to evaluate theoretical aspects of the course. In several faculties this was just a memory exercise consisting in remembering the main theorems' demonstrations, whereas in others this part of the exam was very demanding since students were expected to use the ideas developed in class to demonstrate properties or show counterexamples. These "theoretical exams" have almost disappeared and now the theory is not assessed or—in the best of situations—it is supposed to be evaluated indirectly by asking the learners to say which theorems or properties were used in their resolution of the practical exercises. As a consequence, it is "vox populi" among the students that a theoretical knowledge is not needed in order to do well in the exams. It is only considered useful if the student is trying to earn good marks for any reason (e.g. to obtain a scholarship).

Another important peculiarity, and also a healthy activity, is the existence of private institutes that prepare students for examinations and midterm examinations. For these institutions, Calculus I is perhaps the most important source of funds.

Finally, modeling and applications are not very common in Calculus courses. The situation is similar with the use of technology. In most cases both of them are postponed to second year courses like Differential Equations or Numerical Methods.

2.4.6 Calculus Instruction in Singapore[2]

In Singapore, Calculus instruction begins in upper secondary (Years 9 and 10) as part of the GCE O Level syllabus. Calculus is considered as one of three organizing strands in the content to be covered, viz. Algebra, Geometry and Trigonometry, and Calculus. It takes up about 15 % of curriculum time. Two of four stated objectives of the Additional Mathematics syllabus seem to motivate the learning of calculus at this relatively early stage:

- acquire mathematical concepts and skills for *higher studies* in mathematics and to support learning in the other subjects, in particular, the *sciences*
- connect ideas within mathematics and between mathematics and the sciences through applications of mathematics.

Students who study Physics at A Level require Calculus and, in the overall Singapore mathematics curriculum which relies on a spiral approach to fit in as many useful strands and topics as possible, Calculus has to begin early to support Physics learning. In addition, some students move on to engineering courses in the polytechnics after their O Levels, and the basic calculus that they have learnt is necessary for these courses.

Calculus in Additional Mathematics includes:

- Derivative of f(x) as the gradient of the tangent to the graph of y = f(x) at a point
- Derivative as rate of change
- Derivatives of x^n, sin x, cos x, tan x, e^x, and ln x
- Derivatives of products and quotients of functions
- Derivatives of composite functions
- Increasing and decreasing functions
- Stationary points (maximum and minimum turning points and stationary points of inflexion)
- Use of second derivative test to discriminate between maxima and minima
- Applying differentiation to gradients, tangents and normals, connected rates of change and maxima and minima problems

---

[2]Contributed by Tay Eng Guan, National Institute of Education, Singapore.

- Integration as the reverse of differentiation
- Integration of x^n, sin x, cos x, tan x, $\sec^2 x$, and e^x
- Definite integral as area under a curve
- Finding the area of a region bounded by a curve and line(s)
- Application of differentiation and integration to problems involving displacement, velocity and acceleration of a particle moving in a straight.

The stated objectives of the Additional Mathematics syllabus carry over to the A Level Mathematics syllabus. Calculus at this level includes:

- Sequences, Series and convergence
- Differentiation of simple functions defined implicitly or parametrically
- Locating maximum and minimum points using a graphing calculator
- Finding the approximate value of a derivative at a given point using a graphing calculator
- Finding equations of tangents and normals to curves
- Connected rates of change problems
- Maclaurin series
- Integration of $\frac{f'(x)}{f(x)}$, $\sin^2 x$, $\cos^2 x$, $\tan^2 x$, $\frac{1}{a^2+x^2}$, $\frac{1}{a^2-x^2}$, $\frac{1}{\sqrt{a^2-x^2}}$.
- Integration by a given substitution
- Integration by parts
- Finding the area under a curve defined parametrically
- Finding the volume of revolution about the x- or y-axis
- Finding the approximate value of a definite integral using a graphing calculator
- Solving for the general solutions and particular solutions of differential equations
- Formulating a differential equation from a problem situation
- Interpreting a differential equation and its solution in terms of a problem situation.

The use of a graphing calculator is expected and in fact, is necessary for some assessment items. Graphing calculators are also used to enhance teaching by making graphs, derivatives and areas under curves easier to 'visualize'. Together with linking differential equations to problem situations, graphing calculators are also intended to connect calculus with applications to real life.

Undergraduates taking engineering and science programs in Singapore public universities are expected to have taken Calculus in A Level Mathematics or its equivalent in the polytechnics. Thus, students from overseas (including China) will have to take bridging courses in Calculus before enrolling in engineering and science programs. Calculus courses in undergraduate mathematics programs are intended to lead on to Analysis and more advanced integration courses.

2.4.7 Calculus Instruction in South Korea[3]

Calculus in South Korea is considered to be the most important and essential part of secondary school mathematics. Mathematics teachers and professionals in math-related fields value school calculus not only for its mathematical significance, but also for its wide range of application and strong connection to higher education. As a result, calculus has never been excluded from the Korean Scholastic Aptitude Test (KSAT). This status of calculus makes high school students study calculus up to a certain level, even including those who do not wish to pursue further study in either science or engineering. Sometimes, however, it is the subject of controversy whether all the students must be taught calculus, regardless of further study. Lately there have been discussions about issues in mathematics education: students who give up on studying, low levels of academic satisfaction compared to high level of the achievement, rote computation-oriented learning, intense competition and the pressure for private education in mathematics outside of school.

Korean students mostly begin to learn calculus in the 2nd year (Grade 11) at high school. It is taught in accordance with the current national curriculum, composed in the following order: limit of a sequence, limit of a function, differentiation, integration. These units are presented in two subjects, *Calculus 1* and *Calculus 2*. Most of the Korean high schools provide two different tracks for students, namely the *Liberal Arts* (**LA**) track and the *Natural Sciences* (**NS**) track, although the current curriculum no longer officially stipulates such separation. Students in the LA track usually learn *Calculus 1* whereas those in the NS track learn both *Calculus 1* and *Calculus 2*. Still, students in both tracks go through the same learning sequence. In particular, since a limit is defined without the $\epsilon-\delta$. method, a large portion of the calculus concepts is defined less rigorously and thus taught with a lower level of rigor, whereas the level of difficulty in computations being carried out remains high.

There has been constant revision in the national curriculum, including calculus. The 7th national curriculum, which was announced in 1997 and implemented in 2002 allowed students a choice whether to take the calculus section of the KSAT, which meant that the calculus portion of the college entrance exam could be significantly curtailed, although there were still calculus-related topics.

The following curriculum, the *2007 Revised National Curriculum,* was announced in 2006 and implemented in 2009. Calculus became a mandatory subject for the students in both tracks. LA students learned differentiation and integration of polynomial functions while NS students learned those of the exponential, logarithmic, and trigonometric functions.

The *2009 Revised National Curriculum* is the one announced in 2009 and currently effective since 2011. Along with the emphasis on "mathematical process,"

---

[3]Contributed by Oh Nam Kwon, Seoul National University.

it consists of *Calculus 1*, which covers the calculus of polynomial functions, and *Calculus 2* which does the transcendental functions. Volumes of solids of revolution were moved to the more advanced subject. Those who desire further study can learn advanced application of calculus such as differential equations and partial differentiation in *Advanced Mathematics 2*.

The *2015 Revised National curriculum* was announced in 2015 and is scheduled to be effective by 2018. As the latest mathematical curriculum, it has tried an overall reduction of the amount of content. The Korean Calculus curriculum had defined the definite integral as a limit of a Riemann sum following the sequence of Limit of a sequence, Limit of a function, Differentiation, and Integration, but the new curriculum suggests that we define the definite integral without the limit of a sequence as a response to a critique that a large number of students only compute by rote without understanding:

> Do not cover the definition of the definite integral using the summation of a series. Define a 'definite integral of $f(x)$ from a to b' as '$F(b) - F(a)$' where $F(x)$ is an indefinite integral of $f(x)$, but the introduction and explanation may vary.

It is yet to be discovered whether this new approach to defining the definite integral relieves the students of the difficulty of studying this concept. However, with the constant revision of the calculus curriculum, mathematics educators in South Korea are devoted to responding to public concerns about the difficulties students encounter in the study of calculus and to improving calculus education. Mathematics teachers are developing calculus problems that reflect various real-life-situations and are trying to outgrow the traditional plain lecture class by integrating graphic technology such as graphic calculators, GeoGebra, and GSP. Likewise, South Korean mathematics teachers and researchers are conducting various efforts and research for student-centered mathematics so that students may achieve meaningful learning.

2.4.8 Calculus in Hong Kong[4]

The Hong Kong education system consists of 6 years of primary, 3 years of junior secondary and 3 years of senior secondary. The teaching of calculus starts in the senior secondary section. The mathematics curriculum in the senior secondary consists of a compulsory part and an extended part with two modules. The Mathematics Compulsory Part is meant for all students in Secondary 4–6 (Grade 10–12) while the two modules in the extended part are "designed to cater for students who intend to pursue further studies which require more mathematics; or follow a career in fields such as natural science, computer sciences, technology or engineering." Students with such needs can take at most one of the two modules in the Mathematics Extended Part. These two modules are Module 1 (Calculus

---

[4]Contributed by Wai-man Chu, Ida Ah Chee Mok, and Ka-Lok Wong, University of Hong Kong.

and Statistics) and Module 2 (Algebra and Calculus), in which the instruction of
the Calculus area represents respectively 47.2 and 49.6 % of the curriculum in
terms of the recommended teaching hours. In the public examination of 2015, 5.7
and 8.3 % of all candidates entered the examinations of Module 1 and Module 2
respectively.

Given the concept of function, together with the different representations and
transformation of functions, which should be well understood in the Compulsory
Part, Modules 1 and 2 continue with an intuitive approach to the concept of limit
with no formal definition. The concepts and techniques of differentiation and integration are then introduced gradually.

When compared with Module 1, Module 2 emphasizes mathematical rigour
rather than applications. For example, finding derivatives of functions from first
principles is included in Module 2 but *not* Module 1. In contrast, the trapezoidal
rule is covered in Module 1 but *not* Module 2. The examination may also reveal
the difference in the curricular emphasis. In the public examination of 2015, can-
didates of Module 2 have to evaluate the integrals $\int_1^9 \frac{1}{\sqrt{x}e^{2\sqrt{x}}}dx$, $\int x^2 \ln x\, dx$ and
$\int_4^{10}\left(14 - \frac{x^2+12}{x-2}\right)dx$, while for Module 1, the most complicated integral in the
paper is $\int \frac{t}{1+t}dt$. Moreover, two questions in the Module 1 paper involve the use
of mathematical functions in modeling daily-life situations but there is no such
application problem in Module 2.

The students' performances in the public examination in the Compulsory Part,
Module 1 and Module 2 will be separately reported so as to facilitate the admission to different undergraduate programmes such as engineering, actuarial science
and quantitative finance.

2.5 Calculus Teachers and Classroom Practices

Studies investigating teachers' practices in Calculus usually founded their analysis on the results of questionnaires and interviews. These questionnaires and interviews are built according to the complexity of the underlying Calculus concepts
and potential student difficulties. The common aim is to examine the flexibility
of mainstream teacher practices and how this could affect student learning of
Calculus concepts.

Based on both ATD and TWM frameworks, Smida and Ghedamsi (2006) studied the teaching practices of real analysis in the first year of mathematics courses
in Tunisian universities. They distinguished two kinds of teaching projects: (1) the
projects where axiomatic, structures and formalism are the discourse which justify and generate the expected knowledge; this project only follows a mathematical logic; (2) the projects where the variety of choices for proving, illustrating,
applying or deepening the mathematical results highlights an intent to enroll in a
constructivist setting; this project combines the logic of mathematics and cognitive
demands. A complementary examination of the questionnaire applied to teachers

(lecturers and associate professors) from 4 universities highlights 3 groups of teachers: (1) the teachers with a logico-theoretical profile, who do not take into account cognitive demands; (2) the teachers with a logico-constructivist profile, who have some cognitive concern; (3) the almost one quarter of the teachers who take into account cognitive demands. Furthermore, the great majority of the teachers do not consider the proof in Analysis as a means of convincing students of the validity of mathematical statements, they pointed out the efficiency of founding preliminary analysis courses on numerical methods of approximation in order to give appropriate meanings to Calculus concepts. Drawing on interviews with university teachers, González-Martín et al. (2011) explained how teachers' practices with infinite sums in Québec and UK are related to textbooks expectations. By the means of ATD and RSR frameworks, the analysis pointed out the potentialities shared by teachers of illustrative examples and evocative visual representations in teaching, as well as student engagement with systematic guesswork and writing explanatory accounts of their choices and applications of convergence tests.

Recently, Viirman (2014) studied the teaching practices used by Swedish university mathematics teachers when defining the function concept. Using the COF constructs of *construction and substantiation routines*, the analysis of teacher discourse pointed out subtle differences between practices. According to the researcher, the variation in the construction routines among teachers is related to what he labels construction by stipulation, by exemplar and by contrast. These several types of construction depend on the way in which the teachers underline the mathematical need for the construction of the function.

To investigate teachers' beliefs or goals in Calculus, Eichler and Erens (2014) focused on four systems of beliefs conceived by means of four well known educational trends of teaching Calculus: process-oriented view, application-oriented view, formalist view and schema view. The empirical results show the necessity of distinguishing among central goals, subordinated goals and peripheral goals. According to this study, almost all Calculus teachers have the same peripheral goals related to schema view: Calculus is a set of rules and procedures to be memorized and applied in routine tasks.

Trying to catch as many of the parameters as possible that impact on students' learning, some studies focus on the investigation of classroom evolution when working with Calculus concepts. The theoretical tools that are employed are usually planned in accordance with more than one framework.

To investigate classroom practice on sequence convergence at first-year university, Ghedamsi (2015) designed a methodological tool based on the TDS framework. To tackle the specificities of the transition from secondary school, the definition of the tool was supported by the work of Robert (2007) in the field of teachers' practices. This study suggested a method to illustrate teacher management and its implication on the learning process, as well as a more local description of effective learning on the Calculus concept referred to. In the case of sequence convergence, the analysis highlighted the elements that enabled students' work to shift to university requirement.

In the terms of COF framework, the transition from school to university mathematics requires *substantial discursive shifts*. Building on this, Stadler (2011) used the concept of tangent as a filter to study student-teacher interactions in the transition between secondary and tertiary education in mathematics. This study focused on the differences between school mathematical discourse and scientific mathematical discourse to analyse student difficulties in building bridges between them. Handled on the same assumption, Güçler (2013) explored the discursive shifts experienced in the context of a beginning-level undergraduate Calculus classroom when working with the limit concept. This study strengthened the stages where the teacher implicitly shifts the discourse on limits. For instance, students remained insensitive to shifts relating to discourses on limit as a number and limit as a process.

It is acknowledged that the metaphorical register had a great influence in discussing different definitions. The examination of the metaphorical register in the context of the Calculus classroom has been initiated in some studies. Dawkins (2009) presented a categorization of these metaphor uses (logical metaphor and mathematical metaphor) in an undergraduate real analysis classroom. This study reported that these instructional metaphors lead sometimes to potentially inappropriate conceptions, which are unfortunately integrated into students' images.

Code et al. (2014) conducted a comparative study of Calculus classrooms. Their aim was to investigate the potentialities offered by the interactive engagement teaching model to help students master the conceptual and procedural aspects of the concepts of rate of change and linear approximation. The results revealed that students in the higher engagement classroom were more successful in connecting the procedures to new ideas.

2.6 Current Theoretical Tools for Designing Calculus Tasks

There are several frameworks that have been used to design tasks in the field of Calculus and analysis (COD, TWM, TDS, TAD, etc.). No matter which frameworks are selected, the elaboration of these tasks is conditioned by the learning requirements of the students.

Several design studies related to sequence convergence have been undertaken using the lenses of CID and TWM frameworks. Among these studies, that of Mamona-Downs (2001) provided a series of tasks to encourage students' intuitive images of a sequence as having an ultimate term associated with the limit. This design is based on her analysis of the formal definition of limit via identifying roles for each symbol that occurs to achieve a mental image firmly consonant with the definition. Based on students' images of sequence convergence, Przenioslo (2005) presented a set of specially designed problems and questions for discussion. She argued that this design enables students to develop conceptions that are consonant with the meaning of the concept of limit of a sequence; these conceptions emerge progressively with small jumps between successive stages. Recently, Mamona-Downs (2010) focused on the metaphor of "arbitrary closeness" to propose a design on sequence convergence involving convergence behaviour of a sequence and the accumulation points of the underlying set of the sequence. In the same spirit, Roh (2010b) proposed a hands-on activity, called the ε-strip activity, with a physical device made from translucent paper, as an instructional method to help students construct sequence convergence via visualization of the ε-N definition. Students have to choose the appropriate definition among statements that highlighted the relationship between ε and N. Keene and Hall (2014) used classical Zeno's paradox about geometric sequence to design an alternative for learning the concept of limit. Their aim is to help students to create an informal understanding that modify their intuition to include the concept of arbitrary closeness.

In order to allow first-year university students to perceive the link between real numbers and limit, Ghedamsi (2008) has drawn on TSD constructs to elaborate and experiment with a succession of two situations based on approximation methods. In this study, the link is progressively made through a productive connection of the intuitive, perceptual and formal dimensions of limit. This series of situations endorses an epistemological shift in the students' thought, allowing them to consider real numbers as conceptual objects in close relation to limit within a mathematical theory. The notion of integral has been the meeting point of studies that used epistemological investigation to design situations. González-Martín et al. (2004, 2008) focused on the notion of infinite area to propose an original approach, based on a geometrical setting, to the introduction of improper integral.

Using ATD as framework, Gyöngyösi et al. (2011) described an experiment aimed at using CAS-based work to surmount the difficulties of the transition from Calculus to real analysis. In this study, a set of Calculus praxeologies were designed and analyzed according to their pragmatic value (efficiency of solving tasks) and epistemic value (insight they provide into the mathematical objects and theories to be studied). Job and Schneider (2014) built on the epistemological obstacle called empirical positivism to interpret students' reactions to tasks involving limits. These tasks are related to two kinds of praxeologies: pragmatic and deductive. They argued that a pragmatic praxeological level of rationality should be a preliminary step of development that enables students to perceive several sides of limit concept.

Based on APOS analysis of the concept of infinity, Voskoglou (2013) suggested a didactic approach for teaching real numbers at an elementary level. This approach is designed according to the multiple representations of real number and on the connections between them. Kouropatov and Dreyfus (2013) have used these insights to build and study a curriculum for Israeli high school students that has helped them to construct "integration as a conceptual aggregate of knowledge elements from approximation via accumulation to the FTC." For these researchers, the idea of accumulation is a core concept for a high school integral Calculus curriculum. More recently (Kouropatov and Dreyfus 2014), they particularly focused on the teaching episodes where students deal, for the first time and in an intuitive manner, with the aforementioned notions.

Oehrtman et al. (2014) have demonstrated how encouragement and development of the approximation metaphor, which may indeed be one of Tall's cognitive roots, can be used to help students discover for themselves a mathematically correct definition of limit. In doing so, they have laid a theoretical foundation for an approach to limits that has been used by some textbook authors (Artin 1958; Callahan et al. 2008; Lax and Terrell 2014).

Almost all of these studies and others argued that classical courses exterminate the natural roots of Calculus ideas. The tasks used in these designs are frequently based on historical situations by incorporating the original ideas that allows mathematicians to develop their conceptions of Calculus. Nowadays, intuitions could be modified regarding the new symbolic environment, but the research shows that students' conceptions and historical ideas about Calculus are still firmly intertwined.

2.7 The Use of Technology: A Way for Improving Visualization

As mentioned by Tall et al. (2008), "*of all the areas in mathematics, calculus has received the most interest and investment in the use of Technology*" (p. 207). In this paper, the authors gave a wide range of research related to the role of technology in the teaching and learning of mathematics. The majority of the underlined research highlighted the power of technology to improve visualization skills— namely the skills of forming visual mental images that are in accordance with the desired outcomes, in the first steps of learning mathematics. Later, technology could be used as a programming language to improve mathematical procedures and algorithms.

In the case of Calculus, a visual approach of the fundamental ideas of infinitesimals, approximations process, change, variation, accumulation, etc. via technology may help students to have insights about the formal existence of Calculus concepts (Tall 1986, 1990, 2003, 2013). In the same spirit, Moreno-Armella (2014) claimed that standard analysis does not correctly interrelate the intuition of change and accumulation. He drew on Euler's idea of continuous function as an infinitesimally enriched continuum to emphasis the role of visualization in learning and teaching integration via the use of a mediating artifact—digital media. Arguing that both digital and infinitesimals models are discrete models, he put forward the cognitive relationship between zooming on a graph and taking infinitesimals. The study of Weigand (2014) goes beyond this work by clearly emphasizing the role of the link between digital technology and discrete mathematics in learning about derivatives. They proposed an alternative discrete step-by-step approach to the basic concepts of calculus by developing the concept of rate of change in a discrete learning environment: working with discrete sequences and functions defined on Z and discrete domains of Q. However, the authors carefully underlined the necessity of a conceptual change from the discrete thinking to continuous thinking. For the authors, this alternative should be considered as a transitional situation lying between intuition and formalism.

The Instrumental Genesis Theory (IGT) of Rabardel (1995) strengthened the complexity of the process of transformation of the software used into a mathematical instrument. Using IGT, Henriques (2006) investigated the usefulness of Maple to overcome students' difficulties when calculating areas and volumes by multiple integrals. This study underlines the necessity to deeply investigate the relationship between mathematical knowledge and knowledge about the used software.

Building on the theory of objectification, Swidan and Yerushalmy (2014) explored the ways in which students actively engage in objectifying the concept of indefinite integrals graphically by using dynamic artifacts. In accordance with the objectification theory, the authors considered the artifact as a fundamental part of Calculus thinking, claiming that the role of the teachers and students must be modified.

The role of technology is generally the main theme discussed in the topic study group of learning and teaching Calculus over the last three International Congresses on Mathematical Education (ICME). Most of these contributions investigated the potency of technology by means of empirical perspectives. They generally focus on the interrelation between intuitive and analytic thoughts in teaching and learning the basic ideas of Calculus with mathematical software, including graphic calculators. These ideas are particularly related to the decimal expansions of real numbers and its link with limit notion, the relationship between derivative and integral, multivariable Calculus and so on. Some of these contributions underlined the difficulties in translating these ideas into a digital model in way that maximizes the consistency of students' interpretations. More global instructional approaches were also presented showing a diversity of courses planned by using technology to supplement mathematical learning via applied examples or historical situations.

3 Summary and Looking Ahead

In this chapter, we described the punctual evolution of research that has been approached through the main trends in the field of Calculus education: students' difficulties, classroom practices, task design, etc. A variety of epistemological, cognitive and institutional issues have been raised by this research. One of the most important issues that emerged concerns the dialectical relationship between Calculus students' thinking and institutional expected thinking. There are at least three dialectics that are classically current tensions in this relationship: potential versus actual, dynamic versus static, and visualization versus formalization. The questions are: What epistemological considerations should be taken into account to face such tensions? What is the role of teaching and classroom practices? The ultimate question is: Which Calculus design considerations could associate teacher expectations with students' requirements?

Some of the research presented in this survey put forward several principles for designing Calculus tasks based on both theoretical and empirical points of view. However, this kind of research does not clearly state the impact of their design on students' learning in regular lessons: Do students' interactions influence teacher adjustment of the design? More and more: What is the extent to which teachers' beliefs about learning Calculus impact on the implementation of the design? Does the design interrelate with allocated time in regular lessons?

This survey is not complete; other studies concerning mathematics education without specific reference to Calculus could contribute to extend the research in this field. Among these studies, those that investigate the use of digital technology tools to improve mathematical learning have highlighted the complex process of transforming an artifact into a learning and teaching tool. In the case of Calculus, this kind of research should also deal with the aforementioned dialectics.

These questions and others provide researchers with a fitting basis to move forward toward the goal of improving the research results. In this survey:

- The instructional situation of Calculus through the last twelve years in different parts of the world is analyzed;
- Approaches for an investigation of the institutional Calculus context are described;
- Insight into the main aspects of students' Calculus thinking is described;
- Ideas for designing Calculus tasks are described;
- New research questions about the teaching and learning of Calculus are put forward.

References

Alcock, L., & Simpson, A. (2011). Classification and concept consistency. *Canadian Journal of Science, Mathematics and Technology Education, 11*(2), 91–106.

Arslan, S. (2005). L'approche qualitative des equations différentielles en classe de terminale S: est-elle viable ? Quels sont les enjeux et les conjectures ? (Phd. thesis, Université Joseph-Fourier, Grenoble).

Artin, E. (1958). *A Freshman honors course in calculus and analytic geometry taught at Princeton University.* Buffalo, NY: Committee on the Undergraduate Program of the Mathematical Association of America.

Asiala, M., Cottrill, J., Dubinsky, E., & Schwingendorf, K. E. (2001). The development of students' graphical understanding of the derivative. *Journal of Mathematical Behavior, 16*, 399–431.

Aspinwall, L., Shaw, K. L., & Presmeg, N. C. (1997). Uncontrollable mental imagery: Graphical connections between a function and its derivative. *Educational Studies in Mathematics, 33*, 301–317.

Bachelard, G. (1938). *La formation de l'esprit scientifique.* Paris: Librairie philosophique Vrin.

Bagni, G. T. (2005a). Mathematics education and historical references: Guido Grandi's infinite series. *Nordisk Matematisk Tidsskrift, 53*, 173–185.

Bagni, G. T. (2005b). The historical roots of the limit notion. Cognitive development and development of representation registers. *Canadian Journal of Science, Mathematics and Technology Education, 5*(4), 453–468.

Bagni, G. T. (2007). Didactics and history of numerical series: Grandi, Leibniz and Riccati, 100 years after Ernesto Cesaro's death. *La matematica e la sua didattica, 21*(1), 75–80 (Special Issue).

Baker, B., Cooley, L., & Trigueros, M. (2000). A Calculus graphing schema. *Journal for Research in Mathematics Education, 31*, 557–578.

Bergé, A. (2008). The completeness property of the set of real numbers in the transition from Calculus to analysis. *Educational Studies in Mathematics, 67*(3), 217–236.

Bergé, A. (2010). Students' perceptions of the completeness property of the set of real numbers. *International Journal of Mathematical Education in Science and Technology, 41*(2), 217–227.

Bezuidenhout, J. (2010). Limits and continuity: Some conceptions of first-year students. *International Journal of Mathematical Education in Science and Technology., 32*(4), 487–500.

Bingolbali, E., Monaghan, J., & Roper, T. (2007). Engineering students' conceptions of the derivative and some implications for their mathematical education. *International Journal of Mathematical Education in Science and Technology, 38*(6), 763–777.

Black, M. (1962). Metaphor. In M. Black (Ed.), *Models and metaphors: Studies in language and philosophy* (pp. 219–243). Ithaca, NY: Cornell University Press.

Black, M. (1977). More about metaphor. *Dialectica, 31*, 433–457.

Błaszczyk, P., Katz, M., & Sherry, D. (2013). Ten misconceptions from the history of analysis and their debunking. *Foundations of Science, 18*, 43–74.

Bloch, I., & Ghedamsi, I. (2005). Comment le cursus secondaire prépare-t-il les élèves aux études universitaires? *Petit x, 69*, 7–30.

Borgen, K. L., & Manu, S. S. (2002). What do students really understand? *Journal of Mathematical Behavior, 21*, 151–165.

Borovik, A., & Katz, M. (2012). Who gave you the Cauchy–Weierstrass tale? The dual history of rigorous calculus. *Foundations of Science, 17*(3), 245–276.

Bosch, M., Fonseca, C., & Gascón, J. (2004). Incompletitud de las organizaciones matemáticas locales en las instituciones escolares. *Recherches en Didactique des Mathématiques, 24*(2/3), 205–250.

Bressoud, D. (2015). The calculus student. In D. Bressoud, V. Mesa, & C. Rasmussen (Eds.), *Insights and recommendations from the MAA National Study of College Calculus.* Washington, DC: Mathematical Association of America.

Bressoud, D., Mesa, V., & Rasmussen, C. (2015). Preface. In D. Bressoud, V. Mesa, & C. Rasmussen (Eds.), *Insights and recommendations from the MAA National Study of College Calculus.* Washington, DC: Mathematical Association of America.

Brousseau, G. (1983). Les obstacles épistémologique et les problèmes en mathématiques. *Recherches en Didactique des mathématiques, 4*(2), 165–198.

Brousseau, G. (1997). Theory of didactical situations in mathematics. In N. Balacheff, M. Cooper, R. Sutherland, & V. Warfield (Eds., Trans.).

Byerley, C., Hatfield, N., & Thompson, P. W. (2012). Calculus students' understandings of division and rate. In S. Brown, S. Larsen, K. Marrongelle, & M. Oehrtman (Eds.), *Proceedings of the 15th Annual Conference on Research in Undergraduate Mathematics Education* (pp. 358–363). Portland, OR: SIGMAA/RUME.

Callahan, J., Hoffman, K., Cox, D., O'Shea, D., Pollatsek, H., & Senechal, L. (2008). *Calculus in context: The five college calculus project.* www.math.smith.edu/Local/cicintro/book.pdf. Accessed August 11, 2014.

Carlson, M., Persson, J., & Smith, N. (2003). Developing and connecting Calculus students' notions of rate-of-change and accumulation: The fundamental theorem of Calculus. In *Proceedings of the 2003 Meeting of the International Group for the Psychology of Mathematics Education – North America* (Vol. 2, pp. 165–172). Honolulu, HI: University of Hawaii.

Chellougui, F. (2009). L'utilisation des quantificateurs universel et existentiel en première année d'université: entre l'explicite et l'implicite. *Recherches en Didactique des Mathématiques, 29*(2), 123–154.

Chevallard, Y. (1985). La transposition didactique. La Pensée Sauvage.

Code, W., Piccolo, C., Kohler, D., & MacLean, M. (2014). Teaching methods comparison in a large calculus class. *ZDM Mathematics Education, 46*, 589–601.

College Board. (1997–2014). *AP data–Archived data.* Retrieved August 27, 2015, from College Board Web site: http://research.collegeboard.org/programs/ap/data/archived

College Board. (2015). *AP calculus AB and AP calculus BC curriculum framework 2016–2017.* Retrieved August 27, 2015, from College Board Web site: https://secure-media.collegeboard. org/digitalServices/pdf/ap/ap-Calculus-curriculum-framework.pdf

Confrey, J., & Smith, E. (1994). Exponential functions, rates of change, and the multiplicative unit. *Educational Studies in Mathematics, 26*, 135–164.

Coppé, S., Dorier, J.-L., & Yavuz, I. (2007). De l'usage des tableaux de valeurs et des tableaux de variations dans l'enseignement de la notion de fonction en France en seconde. *Recherche en Didactique des Mathématiques, 27*(2), 151–186.

Cornu, B. (1991). Limits. In D. Tall (Ed.), *Advanced mathematical thinking* (pp. 153–166). Dordrecht, The Netherlands: Kluwer.

Cottrill, J., Dubinsky, E., Nichols, D., Schwingendorf, K., Thomas, K., & Vidakovic, D. (1996). Understanding the limit concept: Beginning with a coordinated process scheme. *Journal of Mathematical Behavior, 15*(2), 167–192.

Czocher, J., Tague, J., & Baker, G. (2013). Coherence from calculus to differential equations. In *Proceedings of the 16th Annual Conference on Research in Undergraduate Mathematics Education* (pp. 283–291).

Davis, R. B., & Vinner, S. (1986). The notion of limit: Some seemingly unavoidable misconception stages. *Journal of Mathematical Behavior, 5*(3), 281–303.

Dawkins, P. (2009). Concrete metaphors in the undergraduate real analysis classroom. In S. L. Swars, D. W. Stinson, & S. Lemons-Smith (Eds.), *Proceedings of the 31st Annual Meeting of the North American Chapter of the International Group for the Psychology of Mathematics Education* (Vol. 5, pp. 819–826).

Dias, M., Artigue, M., Jahn A., & Campos, T. (2008). A comparative study of the secondary tertiary transition. In M. F. Pinto & T. F. Kawasaki (Eds.), *Proceedings of the 34th Conference of the International Group for the Psychology of Mathematics Education* (Vol. 2, pp. 129–136).

Dooley, T. (2009). The development of algebraic reasoning in a whole-class setting. In M. Tzekaki, M. Kaldrimidou, & H. Sakonidis (Eds.), *Proceedings of the 33rd Conference of the International Group for the Psychology of Mathematics Education, Thessaloniki* (Vol. 1).

Dubinsky, E. (1991). Reflective abstraction in advanced mathematical thinking. In Tall, D. (Eds.), *Advanced mathematical thinking* (pp. 95–123).

Dubinsky, E., & McDonald, M. (2001). APOS: A constructivist theory of learning. In D. Holton (Ed.), *The teaching and learning of mathematics at university level: An ICMI study* (pp. 275–282).

Duval, R. (1995). Sémiosis et pensée: registres sémiotiques et apprentissages intellectuels [Semiosis and human thought. Semiotic registers and intellectual learning]. Peter Lang.

Eichler, A., & Erens, R. (2014). Teachers' beliefs towards teaching calculus. *ZDM Mathematics Education, 46*, 647–659.

EMS-Committee of Education. (Eds.). (2014). Solid findings: Concept images in students' mathematical reasoning. *EMS Newsletter, 93*, 50–52.

Ervynk, G. (1981). Conceptual difficulties for first year university students in the acquisition of the notion of limit of a function. In *Proceedings of the Fifth Conference of the International Group for the Psychology of Mathematics Education* (pp. 330–333).

Ferrini-Mundy, J., & Graham, K. (1994). Research in calculus learning: Understanding of limits, derivatives, and integrals. In J. Kaput & E. Dubinsky (Eds.), *Research issues in undergraduate mathematics learning, MAA notes #33*. Washington, DC: Mathematical Association of America.

Ghedamsi, I. (2008). Enseignement du début de l'analyse réelle à l'entrée à l'université : Articuler contrôles pragmatique et formel dans des situations à dimension a-didactique. (PhD Thesis) Université Bordeaux 2, France & Université de Tunis, Tunisie.

Ghedamsi, I. (2015). Teacher management of learning Calculus: The case of sequences at the first year of university mathematics studies. In *Proceedings of CERME9*. To appear.

González-Martín, A. (2009). L'introduction du concept de somme infinie : une première approche à travers l'analyse des manuels. Actes du colloque EMF 2009, (pp. 1048–1061).

González-Martín, A.-S., Bloch, I., Durand-Guerrier, V., & MMaschietto, M. (2014). Didactic situations and didactical engineering in university mathematics: Cases from the study of calculus and proof. *Research in Mathematics Education, 16*(2), 117–134.

González-Martín, A. S., & Camacho, M. (2004). Legitimisation of the graphic register in problem solving at the undergraduate level. The case of the improper integral. In M. Johnsen Høines & A. Berit Fuglestad (Eds.), *Proceedings of the 28th Conference of the International Group for the Psychology of Mathematics Education (PME)* (Vol. 2, pp. 479–486).

González-Martín, A. S., & Correia de Sá, C. (2008). Historical-epistemological dimension of the improper integral as a guide for new teaching practices. In E. Barbin, N. Stehlikova, & C. Tzanakis (Eds.), *History and epistemology in mathematics education: Proceedings of the 5th European summer university* (pp. 211–223).

González-Martín, A. S., Nardi, E., & Biza, I. (2011). Conceptually-driven and visually-rich tasks in texts and teaching practice: The case of infinite series. *International Journal of Mathematical Education in Science and Technology, 42*(5), 565–589.

Grundmeier, T. A., Hansen, J., & Sousa, E. (2006). An exploration of definition and procedural fluency in integral calculus. *PRIMUS, 16*, 178–191.

Güçler, B. (2013). Examining the discourse on the limit concept in a beginning-level calculus classroom. *Educational Studies in Mathematics, 82*(3), 439–453.

Gyöngyösi, E., Solovej, J. P., & Winsløw, C. (2011). Using CAS based work to ease the transition from calculus to real analysis. In M. Pytlak, T. Rowland, & E. Swoboda (Eds.), *Proceedings of CERME 7* (pp. 2002–2011).

Habre, S., & Abboud, M. (2006). Students' conceptual understanding of a function and its derivative in an experimental Calculus course. *The Journal of Mathematical Behavior, 25*, 57–72.

Haddad, S. (2013). Que retiennent les nouveaux bacheliers de la notion d'intégrale enseignée au lycée? *Petit x, 92*, 7–32.

Hardy, N. (2009). Students' perceptions of institutional practices: The case of limits of functions in college level calculus courses. *Educational Studies in Mathematics, 72*(3), 341–358.

Henriques, A. (2006). L'enseignement et l'apprentissage des intégrales multiples: Analyse didactique intégrant l'usage du logiciel Maple. (PhD Thesis) Université Joseph-Fourier, Grenoble.

Job, P., & Schneider, M. (2014). Empirical positivism, an epistemological obstacle in the learning of calculus. *ZDM Mathematics Education, 46*, 635–646.

Keene, K. A., Hall, W., & Duca, A. (2014). Sequence limits in calculus: Using design research and building on intuition to support instruction. *ZDM Mathematics Education, 46*, 561–574.

Kidron, I., & Tall, D. (2015). The role of embodiment and symbolism in the potential and actual infinity of the limit process. *Educational Studies in Mathematics* (8, 2014).

Kim, D.-J., Ferrini-Mundy, J., & Sfard, A. (2012). How does language impact the learning of mathematics? Comparison of English and Korean speaking university students' discourses on infinity. *International Journal of Educational Research, 51–52*, 86–108.

Kirsch, A. (1976). Eine intellektuell ehrliche Einführung des Integralbegriffes in Grundkursen. *Didaktik der Mathematik, 4*, 97–105.

Kouropatov, A., & Dreyfus, T. (2013). Constructing the fundamental theorem of calculus. In A. M. Lindmeier & A. Heinze (Eds.), *Prceedings of the 37th Conference of the International Group for the Psychology of Mathematics Education* (Vol. 3, pp. 201–208). Kiel, Germany: PME.

Kouropatov, A., & Dreyfus, T. (2014). Learning the integral concept by constructing knowledge about accumulation. *ZDM Mathematics Education, 46*, 33–548.

Kuzniak, A., Montoya, E., Vandebrouck, F., & Vivier, L. (2015). *Le travail mathématique en Analyse de la fin du secondaire au début du supérieur: identification et construction*. La pensée sauvage: Ecole d'été de didactique des mathématiques.

Lax, P., & Terrell, M. S. (2014). *Calculus with applications* (2nd ed.). New York, NY: Springer.

Mamona-Downs, J. (2001). Letting the intuitive bear on the formal; a didactical approach for the understanding of the limit of a sequence. *Educational Studies in Mathematics, 48*, 259–288.

Mamona-Downs, J. (2010). On introducing a set perspective in the learning of limits of real sequences. *International Journal of Mathematical Education in Science and Technology, 41*(2), 277–291.

Martin, J. (2013). Differences between experts' and students' conceptual images of the mathematical structure of Taylor series convergence. *Educational Studies in Mathematics, 82*, 267-283.

Martínez-Sierra, G. (2008). From the analysis of the articulation of the trigonometric functions to the corpus of Eulerian analysis to the interpretation of the conceptual breaks present in its scholar structure. In *Proceedings of the HPM 2008 Conference, History and Pedagogy of Mathematics*.

Moreno-Armella, L. (2014). An essential tension in mathematics education. *ZDM Mathematics Education, 46*, 621–633.

Nair, G. S. (2010). *College students' concept images of asymptotes, limits, and continuity of rational functions*. Unpublished doctoral dissertation, Ohio State University, Columbus.

Nardi, E., Ryve, A., Stadler, E., & Viirman, O. (2014). Commognitive analyses of the learning and teaching of mathematics at university level: The case of discursive shifts in the study of calculus. *Research in Mathematics Education, 16*(2), 182–198.

Nemirovsky, R., & Rubin, A. (1992). *Students' tendency to assume resemblances between a function and its derivative*. TERC working paper, Cambridge, MA (pp. 2–92).

Oehrtman, M. (2003). Strong and weak metaphors for limits. In N. Pateman, B. Dougherty, & J. Zilliox (Eds.), *Proceedings of the 27th Conference of the International Group for the Psychology of Mathematics, Honolulu, HI* (Vol. 3, pp. 397–404).

Oehrtman, M. (2009). Collapsing Dimensions, physical limitation, and other student metaphors for limit concepts. *Journal For Research In Mathematics Education, 40*(4), 396–426.

Oehrtman, M., Swinyard, C., & Martin, J. (2014). Problems and solutions in students' reinvention of a definition for sequence convergence. *Journal of Mathematical Behavior, 33*, 131–148.

Orton, A. (1980). *A cross-sectional study of the understanding of elementary Calculus in adolescents and young adults*. Unpublished Ph.D. thesis, University of Leeds.

Orton, A. (1983a). Students' understanding of integration. *Educational Studies in Mathematics, 14*, 1–18.

Orton, A. (1983b). Students' understanding of differentiation. *Educational Studies in Mathematics, 14*, 235–250.

Park, J. (2013). Is the derivative a function? If so, how do students talk about it? *International Journal of Mathematical Education in Science and Technology, 44*(5), 624–640.

Pettersson, K., & Scheja, M. (2008). Algorithmic contexts and learning potentiality: A case study of students' understanding of calculus. *International Journal of Mathematical Education in Science and Technology, 39*(6), 767–784.

Praslon, F. (2000). Continuites et ruptures dans la transition Terminale S/DEUG Sciences en analyse. Le cas de la notion de dérivée et son environnement. In T. Assude & B. Grugeon (Eds.), *Actes du Séminaire National de Didactique des Mathématiques* (pp. 185–220).

Przenioslo, M. (2004). Images of the limit of function formed in the course of mathematical studies at the university. *Educational Studies in Mathematics, 55*(1/3), 103–132.

Przenioslo, M. (2005). Introducing the concept of convergence of a sequence in secondary school. *Educational Studies in Mathematic, 60*(1), 71–93.

Rabardel, P. (1995). Les hommes et les technologies - Approche cognitive des instruments contemporains, Editions Armand Colin.

Rasslan, S., & Tall, D. (2002). Definitions and images for the definite integral concept. In A. Cockburn & E. Nardi (Eds.), *Proceedings of the 26th Conference of the International Group for the Psychology of Mathematics Education, Norwich, UK*.

Robert, A. (1982). L'acquisition de la notion de convergence des suites numeriques dans l'enseignement superieur. *Recherches en Didactiques des Mathematiques, 3*, 307–341.

Robert, A. (2007). Stabilité des pratiques des enseignants de mathématiques (second degré): une hypothèse, des inférences en formation. *Recherches en Didactique des Mathématiques, 27*(3), 271–310.

Roh, K. (2008). Students' images and their understanding of definitions of the limit of a sequence. *Educational Studies In Mathematics, 69*(3), 217–233.

Roh, K. (2010a). An empirical study of students' understanding of a logical structure in the definition of the limit of a sequence via the ε-strip activity. *Educational Studies in Mathematics, 73*, 263–279.

Roh, K. (2010b). How to help students conceptualize the rigorous definition of the limit of a sequence. *Problems, Resources, and Issues in Mathematics Undergraduate Studies, 20*(6), 473–487.

Sealey, V. (2006). Definite integrals, Riemann sums, and area under a curve: What is necessary and sufficient? In S. Alatorre, J. L. Cortina, M. Sáiz, & A. Méndez (Eds.), *Proceedings of the 28th annual meeting of the North American Chapter of the International Group for the Psychology of Mathematics Education*. Mérida, México: Universidad Pedagógica Nacional.

Sealey, V., & Oehrtman, M. (2007). Calculus students' assimilation of the Riemann integral into a preciously established limit structure. In T. Lamberg & L. Wiest (Eds.), *Proceedings of the 29th annual meeting of the North American chapter of the International Group for the Psychology of Mathematics Education. Stateline*. NV: University of Nevada.

Sfard, A. (1991). On the dual nature of mathematical conceptions: Reflections on process and objects as different sides of the same coin. *Educational Studies in Mathematics, 22*, 1–36.

Sfard, A. (2008). *Thinking as communicating: Human development, the growth of discourses, and mathematizing*. New York: Cambridge University Press.

Sierpinska, A. (1985). Obstacles épistémologiques relatifs à la notion de limite. *Recherche en Didactique des Mathématiques, 6*(1), 5–67.

Sierpinska, A. (1987). Humanities students and epistemological obstacles relating to limits. *Educational Studies in Mathematics, 18*, 371–397.

Sierpinska, A. (1990). Some remarks on understanding in mathematics. *For the Learning of Mathematics, 10*(3), 24–36.

Smida, H., & Ghedamsi, I. (2006). Pratiques enseignantes dans la transition lycée/université en analyse. In N. Bednarz (Ed.), *Actes du Colloque EMF 2006*, CD-ROM (pp. 1–24).

Sofronas, K. S., DeFranco, T. C., Vinsonhaler, C., Gorgievski, N., Schroeder, L., & Hamelin, C. (2011). What does it mean for a student to understand the first-year calculus? Perspectives of 24 experts. *The Journal of Mathematical Behavior, 30*, 131–148.

Stadler, E. (2011). The same but different—Novice university students solve a textbook exercise. In M. Pytlak, T. Rowland, & E. Swoboda (Eds.), *Proceedings of CERME7* (pp. 2083–2092).

Steen, L. A., & Dossey, J. A. (1986). Letter endorsed by the governing boards of the Mathematical Association of America and the National Council of Teachers of Mathematics concerning Calculus in the secondary schools. Reprinted in J. P. Gollub, M. W. Bertenthal, J. B. Labov, & P. C. Curtis (Eds.) (2002), *Learning and Understanding: Improving advanced study of mathematics and science in U.S. high schools.* Washington, DC: National Academy Press.

Swidan, O., & Yerushalmy, M. (2014). Learning the indefinite integral in a dynamic and interactive technological environment. *ZDM Mathematics Education, 46*, 517–531.

Swinyard, C. (2011). Reinventing the formal definition of limit: The case of Amy and Mike. *The Journal of Mathematical Behavior, 30*(2), 93–114.

Swinyard, C., & Larsen, S. (2012). Coming to understand the formal definition of limit: Insights gained from engaging students in reinvention. *Journal For Research In Mathematics Education., 43*(4), 465–493.

Szydlik, J. E. (2000). Mathematical beliefs and conceptual understanding of the limit of a function. *Journal for Research in Mathematics Education, 31*(3), 258–276.

Tall, D. (1986). A graphical approach to integration and the fundamental theorem. *Mathematics Teaching, 113*, 48–51.

Tall, D. (1990). Using computer environments to conceptualize mathematical ideas. In *Proceedings of Conference on New Technological Tools in Education, Nee Ann Polytechnic, Singapore* (pp. 55–75).

Tall, D. (1992). The transition to advanced mathematical thinking: Functions, limits, infinity, and proof. In D. Grouws (Ed.), *Handbook of research on mathematics teaching and learning* (pp. 495–511). New York: Macmillan.

Tall, D. (2003). Using technology to support an embodied approach to learning concepts in mathematics. In L. Carvalho & L. Guimara˜es (Eds.), *Historia e tecnologia no ensino da matematica* (Vol. 1, pp. 1–28).

Tall, D. (2004). The three worlds of mathematics. *For the Learning of Mathematics, 23*(3), 29–33.

Tall, D. (2013). A sensible approach to the calculus. F. Pluvinage & A. Cuevas (Eds.), *Handbook on calculus and its teaching.* Mexico: Pearson.

Tall, D. O. (2008). The transition to formal thinking in mathematics. *Mathematics Education Research Journal, 20*(2), 5–24.

Tall, D., & Katz, M. (2014). A cognitive analysis of Cauchy's conceptions of function, continuity, limit, and infinitesimal, with implications for teaching the calculus. *Educational Studies in Mathematics, 86*(1), 97–124.

Tall, D., & Schwarzenberger, R. L. (1978). Conflicts in the learning of real numbers and limits. *Mathematics Teaching, 82*, 44–49.

Tall, D., Smith D., & Piez, C. (2008). Technology and calculus. In M. K. Heid & G. M. Blume (Eds), *Research on technology and the teaching and learning of mathematics* (Vol. 1, pp. 207–258).

Tall, D., & Vinner, S. (1981). Concept image and concept definition in mathematics with particular reference to limits and continuity. *Educational studies in mathematics, 12*(2), 151–169.

Tallman, M., Carlson, M. P. Bressoud, D., & Pearson, M. (2016). A characterization of calculus I final exams in U.S. colleges and universities. *International Journal of Research in Undergraduate Mathematics Education, 2*(1), 105–133.

Thompson, P. W. (1994). Images of rate and operational understanding of the fundamental theorem of Calculus. *Educational Studies in Mathematics, 26*, 229–274.

Thompson, P. W. (1995). Students, functions, and the undergraduate curriculum. In E. Dubinsky, A. H. Schoenfeld, & J. Kaput (Eds.), *Research in collegiate mathematics education I* (pp. 21–44). Providence, RI: American Mathematical Society.

Thompson, P. W., & Carlson, M. (in press). Variation, covariation, and functions: Foundational ways of mathematical thinking. To appear in J. Cai (Ed.), *Third handbook of research in mathematics education*. Reston, VA: National Council of Teachers of Mathematics.

Thompson, P. W., & Silverman, J. (2008). The concept of accumulation in calculus. In M. P. Carlson & C. Rasmussen (Eds.), *Making the connection: Research and teaching in undergraduate mathematics education* (pp. 43–52). MAA Notes #73. Washington, DC: Mathematical Association of America.

Törner, G., Potari, D., & Theodossios, Z. (2014). Calculus in European classrooms: Curriculum and teaching in different educational and cultural contexts. *ZDM Mathematics Education, 46,* 549–560.

Trigueros, M., & Martínez-Planell, R. (2015). Two-variable functions: Analysis from the point of view of a dialogue between APOS theory and ATD. *Ensenanza de las Ciencias, 33*, 157–171.

Vandebrouck, F. (2011). Perspectives et domaines de travail pour l'étude des fonctions. *Annales de Didactique et de Sciences Cognitives., 16*, 149–185.

Viirman, O. (2014). The functions of function discourse—University mathematics teaching from a commognitive standpoint. *International Journal of Mathematical Education in Science and Technology, 45*(4), 512–527.

Vinner, S. (1991). The role of definitions in teaching and learning mathematics. In D. O. Tall (Ed.), *Advanced mathematical thinking* (pp. 65–81). Dordrecht: Kluwer.

Vinner, S., & Hershkowitz, R. (1980). Concept images and common cognitive paths in the development of some simple geometrical concepts. In R. Karplus (Ed.), *Proceedings of the 4th International Conference of the International Group for the Psychology of Mathematics Education (PME)* (pp. 177–184). Berkeley (CA): University of California, Lawrence Hall of science.

Voskoglou, M. (2013). An Application of the APOS/ACE approach in teaching the irrational numbers. *Journal of Mathematical Sciences & Mathematics Education, 8*(1), 30–47.

Warren, E. (2005). Young Children's ability to generalise the pattern rule for growing patterns. In H. L. Chick & J. L. Vincent (Eds.), *Proceedings of the 29th Conference of the International Group for the Psychology of Mathematics Education* (Vol. 4, pp. 305–312).

Warren, E., Miller, J., & Cooper, T. J. (2013). Exploring young students' functional thinking. *PNA, 7*(2), 75–84.

Weigand, H.-G. (2014). A discrete approach to the concept of derivative. *ZDM Mathematics Education, 46,* 603–619.

White, P., & Mitchelmore, M. (1996). Conceptual knowledge in introductory Calculus. *Journal for Research in Mathematics Education, 27*, 79–95.

Williams, S. R. (1991). Models of limit held by college calculus students. *Journal for research in Mathematics Education, 22*(3), 219–236.

Winsløw, C. (2008). Transformer la théorie en tâches: La transition du concret à l'abstrait en analyse réelle. In A. Rouchier, et al. (Eds.), *Actes de la XIIIième école d'été de didactique des mathématiques.*

Winsløw, C., Barquero, B., De Vleeschouwer, M., & Hardy, N. (2014). An institutional approach to university mathematics education: From dual vector spaces to questioning the world. *Research in Mathematics Education, 16*(2), 95–111.

Zandieh, M. (2000). A theoretical framework for analyzing student understanding of the concept of derivative. In E. Dubinsky, A. H. Schoenfeld, & J. Kaput (Eds.), *Research in Collegiate Mathematics Education* (Vol. IV, pp. 103–127). Providence, RI: American Mathematical Society.